I0488365

3rd Grade Science Volume 2

© 2013 Todd Deluca
OnBoard Academics, Inc
Newburyport, MA 01950

800-596-3175

www.onboardacademics.com

©2014 OnBoard Academics, Inc. ALL RIGHTS RESERVED. This book contains material protected under International and Federal Copyright Laws and Treaties. Any unauthorized reprint or use of this material is prohibited. No part of this book may be reproduced or transmitted in any form or by any means, electronic or mechanical, including photocopying, recording, or by any information storage and retrieval system without express written permission from the author / publisher. The author grants teacher the right to print copies for their students. This is limited to students that the teacher teachers directly. This permission to print is strictly limited and under no circumstances can copies may be made for use by other teachers, parents or persons who are not students of the book's owner.

Table of Contents

Fossils

Who is walking across the snow?

You make footprints when you walk your dog in the snow. A footprint shows where you've been, but it will disappear when there is more snow or when the snow melts. Fossils are a little bit like footprints, but they can tell us about animals and plants that existed millions of years in the past.

Fossils are the ancient remains of animals or plants preserved in the earth.

Identify these fossils by labeling them. Can you guess the animal that they come from?

A fossil often looks different from the animal. See if you can label these fossils correctly now that you can connect them to the living animal's image.

How does an animal turn into a fossil.

Think about a fish swimming in the ocean millions of years ago. The fish is coming to the end of a long and happy fish life during which it has avoided the many fierce predators in the ocean at that time. When the fish finally dies of old age, the fish sinks peacefully to the bottom of the ocean.

Over time, bits and sand and rock accumulate on top of the fish and bury it below the ocean floor. Then, as the fish's body begins to rot, tiny bits of sand and rock replace the rotted body parts. This is one way that fossils are formed and why fossils sometimes look like animals that have been turned into stone.

The fossil's flat appearance is due to the weight of all the rock and sand above it.

Fossils are sometimes discovered as a result of earthquakes or volcanoes which can force rocks up to the Earth's surface. Rocks that make it to the surface are great teachers if they contain fossils because they can show us what animals looked like millions or even billions of years ago. Sometimes the fossils offer amazing detail. This is really helpful because most of the animals in these fossils are extinct which means their species is no longer living

Fossils are found in rocks and are normally millions or even billions of years old. There are a number of different ways that animals can turn into fossils after they die.

Fossils can also come from plants.

Match the fossil with the living plants' image.

Match the prehistoric fossil with its living relative of today.

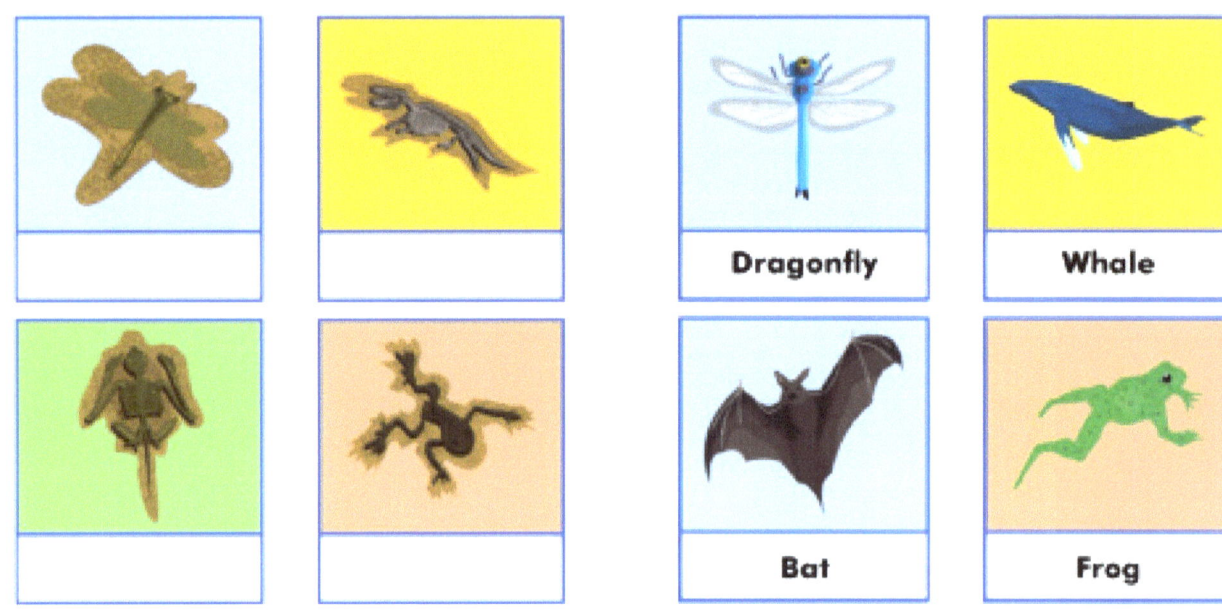

Fossil records show us that many animals looked quite different in the past. For example, dragonflies and frogs were much larger than they are today and whales were much smaller than they are today.

Fossils can show us how the Earth itself has changed.

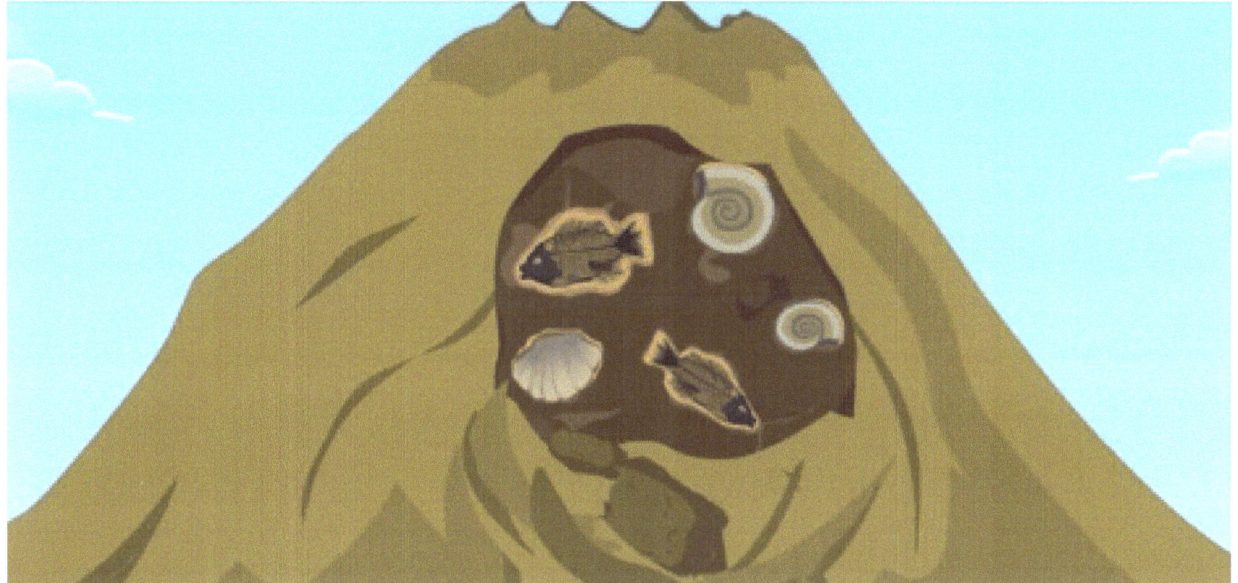

Fish and sea shells on the top of a mountain? How did this happen? Fossils show us that Earth has changed a lot over time. For example, land that was once deep under oceans is now high up in mountains. Fossils really are great teachers of history!

Fossils Quiz

√ for Fact   X for Fiction

Fossils come from animals and plants	
Fossils are found in the ground	
Fossils are identical to the animals they come from	
Fossils can teach us about Earth's history	
Plants never become fossils	
Fossils usually take about 10 years to form	
Fossils are hard	

Adaptation

Traits you inherit from your parents are passed on in the form of genes. A single gene or

a combination of many genes are what account for inherited traits like the shape of your nose or your eye color. Genes are the reason that you may look like other members of your family and are the reason you have your own unique characteristics.

Inherited traits often give an animal an advantage within its environment. For example the long neck of the giraffe enables it to feed at a higher level than all the other animals in its environment. Its long neck also helps it to spot predators from long distances. Traits that help an animal to survive in its environment are called adaptations. Other common examples are webbed feet, sharp teeth, acute vision or thick fur.

But, the giraffe did not always have a long neck. Fossil records suggest that a giraffe's ancestors looked like a cross between a deer and a horse. The adaptation of the giraffe that enabled it to feed more successfully in its environment occurred over a period of millions of years.

But how did this happen? Adaptations occur as a result of a process called natural selection.

To understand natural selection, lets imagine we have a family of green lizards who live in a desert habitat.

Two of the lizards, Fred and Bert, are best friends and hang out together all the time. But, Bert isn't like all the other lizards. His skin is brown and coincidentally the same color as the rocks that the lizards are relaxing on.

Unfortunately Fred and Bert's friendship comes to a sudden end when Fred was carried off by a hungry eagle. Fred was easy to spot because his green skin color stood out against the rock. Bert survived because his his skin color acted as camouflage against the rock.

Since Fred was snatched by the hungry eagle he will be unable to breed, and so in the future there will be fewer green lizards in this environment.

Since Bert was hidden from the eagle by his skin color he will be around for a longer time and more likely to breed and pass on his trait of a brown skin color to his offspring.

That's adaptation in a nutshell. Animals who inherit traits that help their survival are more likely to be around and to breed and pass along those characteristics to their future generations.

Over time the process of natural selection reenforces the continuation of characteristics that help an animal to survive and thrive in its environment until the whole population has those characteristics.

The adaptations that organisms make are directly linked to the environment in which they live. Adaptations help species to survive if their environment changes. Unlike our lizard friends, adaptations don't take place quickly but is a process that happens over many, many generations with many small changes over many millions of years.

What do you notice about the first illustration of the lizard family compared to the second?

_____

_____

Why did the eagle pick Fred instead of Bert? _____

_____

List two reasons that the giraffe adapted its neck size.

1._____

2._____

How did you parents pass traits onto you? _____

_____

Connect the animal with the described adaptation.

My spotted fur helps me to hide in the rain forest. I have powerful jaws for breaking turtle shells. I am short, stocky and a great swimmer.

My chiseled teeth are great for cutting down trees. My webbed feet and flat tail help me to move through rivers and ponds.

My wide flat leaves and feathery roots help me to float on the water.

My powerful legs are great for jumping and swimming. I can catch insects with my sticky tongue. I survive winter by hibernating.

How is my beak adapted to my habitat and diet?

Connect the bird with correct description.

I scoop up fish with my elastic throat pouch.

My long curved beak helps me to extract nectar from flowers.

Hairy structures in my odd shaped beak help me to filter out mud and algae from the shrimp and fish that I eat.

I use my thick, chiseled like beak to extract insects and sap.

Behavior Adaptation

When an animal develops a trait to survive and thrive in its environment we call this a structural adaptation. When an organism changes what it does to survive and thrive in an environment we call this a behavioral adaptation.

For example, when you put on a coat and hat in the winter this is called a behavioral adaptation. You are adapting to colder weather.

Animals and plants also adapt their behavior to changes in their environment. This is particularly true to seasonal changes when foods become scarce. A fox will change its diet throughout the year. In spring and summer the fox eats lots of berries, grasses and insects but in the winter it eats mammals like moles, mice and squirrels.

Hibernation in another adaptation that is motivated by seasonal changes and the availability of food. Animals like chipmunks, snakes and some bears go into a deep sleep during the winter when food is no longer available and awaken in the spring when food is available again.

Some animals migrate, they move to another location, in order to find food and water. Canadian Geese migrate south during the winter because the ponds and other waterways that they rely on for food freeze in the winter.

Plants adapt to seasonal changes too. Some trees shed their leaves during the winter to conserve energy and retain water. Plants also adapt to their environment in other ways. Below ground, the roots of a plant will grow toward the source of water while above ground the plant will grow toward the source of sunlight.   Sometimes a plant will attach itself to a wall or fence to help supporting itself away from the ground.
We call this plant behavioral adaptation tropism.

Many behavioral adaptations are instinctive. For example, when baby turtles hatch they instinctively head toward the water as soon as possible in order to avoid being picked on by predators. Other behavior adaptations can be learned. For example when we train domestic dogs, our pets, to do tricks for treats.

©2013 OnBoard Academics, Inc.

Identify if the adaption is structural or behavioral.

S for structural adaptation

B for behavioral adaption

1. The fur of the snowshoe hair changes from short and brown in the summer to thick, long and white in the winter. _____

2. When threatened opossums play dead, mimicking the smell and appearance of a dead animal. _____

3. Some species of squirrels bury nuts in many different areas. This is called scatter hoarding and prevents a squirrels entire stash from being discovered by another squirrel. _____

4. A porcupine has sharp stiff quills that act as a defense agains predators. _____

5. Some harmless milk snakes look like the very poisonous coral snake. This is called mimicry and helps to deter predators._____

6. Coyotes and badgers help each other to hunt and trap prey. Badgers do the digging and coyotes do the chasing. _____

Adaptation Quiz

1. Traits that help an animal survive in its environment are called _____.
   a. habitats
   b. adaptations
   c. characteristics
   d. genes

2. Adaptations happen over a course of a few generations. True or false?

3. Traits inherited by us from our parents are passed on in the form of _____.
   a. habitats
   b. adaptations
   c. characteristics
   d. genes

4. When an organism develops a physical trait in order to survive in an environment, we call this a _____.
   a. structural adaptation
   b. behavioral adaptation

5. A porcupine's quill is an example of a _____.
   a. structural adaptation
   b. behavioral adaptation

Dinosaurs

Dinosaurs lived long ago.

The period when dinosaurs were alive is from about 230 million years ago to about 65 million years ago. Dinosaurs then became extinct.

Here are some dinosaur facts. Can you tell which are true and which are false?

√ for true
X for false

All dinosaurs were reptiles ☐

All dinosaurs were large ☐

All dinosaurs ate meat ☐

All dinosaurs had scales ☐

All dinosaurs laid eggs ☐

All dinosaurs were slow ☐

There were over 1,000 different types of dinosaurs. Can you select the correct name for these popular six dinosaurs?

Apatosaurus   **Triceratops**   **Tyrannosaurus Rex**

Ankylosaurus   **Stegosaurus**   **Velociraptor**

answer

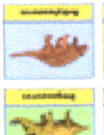 

Can you place these dinosaurs on the line from smallest to largest?

SMALLEST ———————————————————————————→ BIGGEST

TYRANNOSAURUS REX   APATOSAURUS   TRICERATOPS   VELOCIRAPTOR

answer

Some dinosaurs ate meat and some just plants.

TRICERATOPS

VELOCIRAPTOR

YRANNOSAURUS REX

APATOSAURUS

ANKYLOSAURUS

STEGOSAURUS

How did scientists discover dinosaurs? _____

_____

Dinosaurs lived millions of years ago, long before there were people, and so no-one has actually ever seen a living dinosaur. However, paleontologists (scientists who study animals and plants that lived in the past) have discovered lots of fossilized dinosaur bones in old rocks all over the world.

Where did the dinosaurs go?

Why are there no dinosaurs around today?

Scientists have a few ideas about what caused the dinosaurs to become extinct.

The most popular idea is that a very large asteroid from outer space smashed into the earth causing a massive, think blanket dust cloud that covered the Earth and blocked the rays of the sun. This caused a massive extinction of dinosaurs and plants.

Another theory is that a very large volcano erupted cause a dust and ash cloud that blocked the rays of the sun.

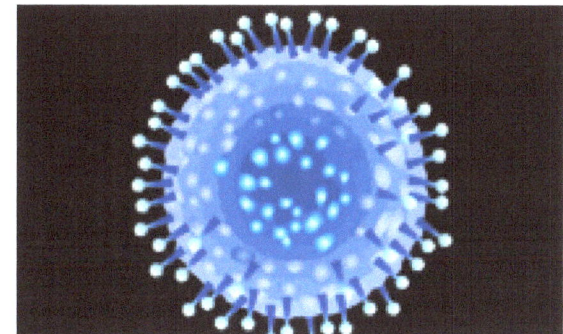

Other idea are a change in the climate on earth or a global virus that infected the dinosaurs. Of course it might have been more than one factor that caused the extinction.

Whatever caused the dinosaurs' extinction the fossil records suggest that it happened quite quickly.

Dinosaurs Quiz

1. Dinosaurs lived about 65 million years ago. True or false?

2. All dinosaurs were large and slow. True or false?

3. There were over _____ different types of dinosaurs.
   - a. 10
   - b. 100
   - c. 1000

4. Ankylosaurus is a plant eating dinosaur. True or false?

5. _____ was a meat eating dinosaur.
   - a. Triceratops
   - b. Velociraptor

Animal Life Cycle

Match each animal with its infant.

What is a life cycle.

A life cycle is the name we give to the set of stages that t a living creature goes through.

When human beings are born, we start out as babies, then we grow and develop into toddlers and then a bit later into kids. At the age of 13 we are officially teenagers and then become adults. We are adults for the longest period in our life cycle. After adulthood we become seniors.

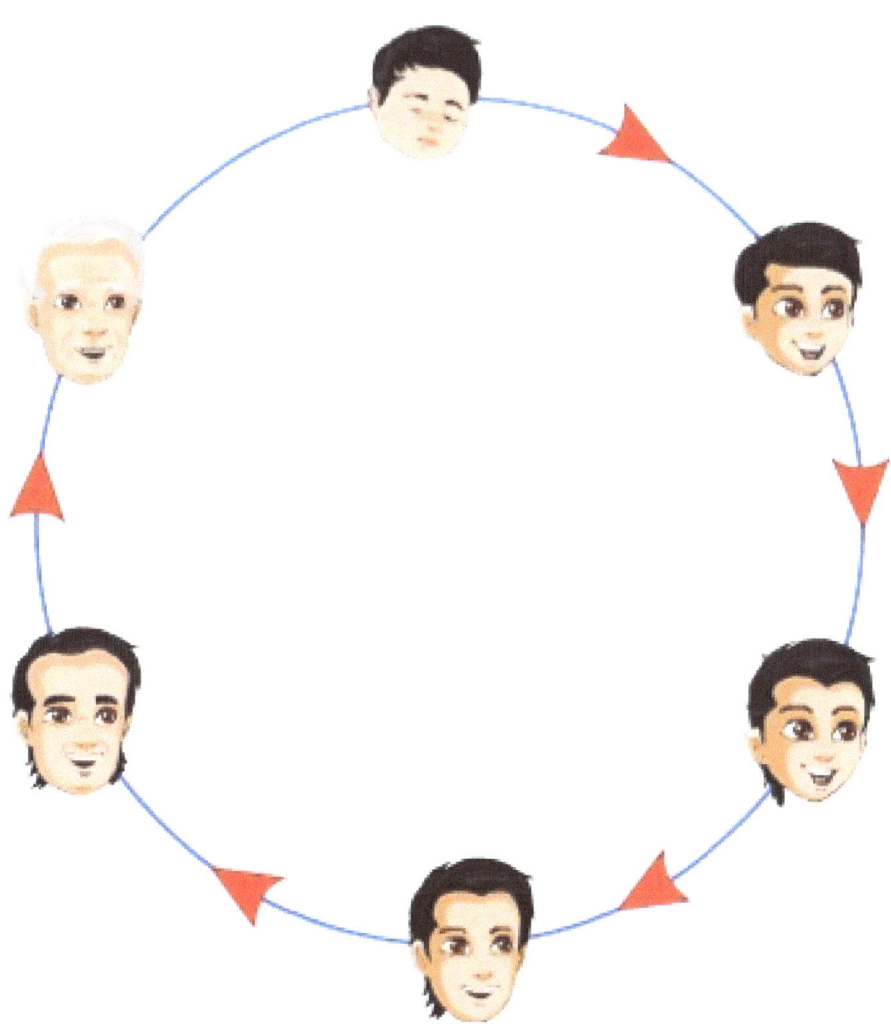

Life cycles of a chicken, a dog and a dragonfly.

 All birds including chickens start their life as an embryo within an egg. The embryo take about 21 days to grow into a chick and hatch. The yolk of an egg is the food source and the hard outer shell of the egg protects the embryo. The chick uses its beak to break out of the shell and sometimes it takes a whole day. The chicks first feathers are down and it takes about 4 weeks to grow its outdoor feathers. It takes about 6 months for the chick to grow into an adult chicken. Female chickens are called hens and male chickens are called roosters.

| Egg | Embryo | Chick | Hen | Rooster |

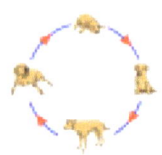 A puppy is born in a litter often with about half a dozen other puppies. Newborn puppies are born with their eyes close and don't normally open their eyes for 2 weeks. During the first few months the puppy will grow quickly and become bigger and heavier. After six weeks it will no longer need its mother's milk. At about two years the puppy will be a fully grown dog and continue the life cycle by having its own puppies although dogs can have puppies at just one year old. As dogs age, the hair around their mouth often turns white or gray. Dogs usually live about 12 years but some will live longer.

©2013 OnBoard Academics, Inc.   www.onboardacademics.com   35

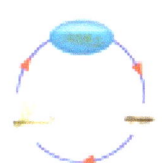 All dragonflies start out as eggs. The mother dragon fly lays the eggs in the water. Some of the eggs will hatch in about 5 days but others will take months to hatch. At this stage the dragonfly is a larva or nymph which is the second stage of a dragon fly's life cycle. This is the longest stage of a dragonflies life spent entirely under water and can last up to four year. Eventually the larva grows wings and turns into a dragon fly and no longer lives under water. The adult stage of a dragon fly's life is quite brief lasting only a couple of months. During this time the dragon fly spends it time looking for a mate to continue the life cycle.

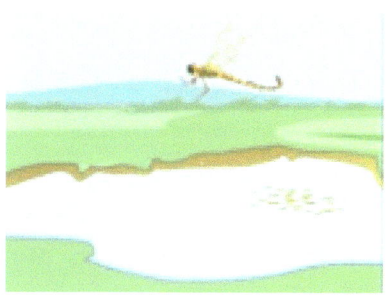

Complete this dog's lifecycle with the suggestions and illustrations below.

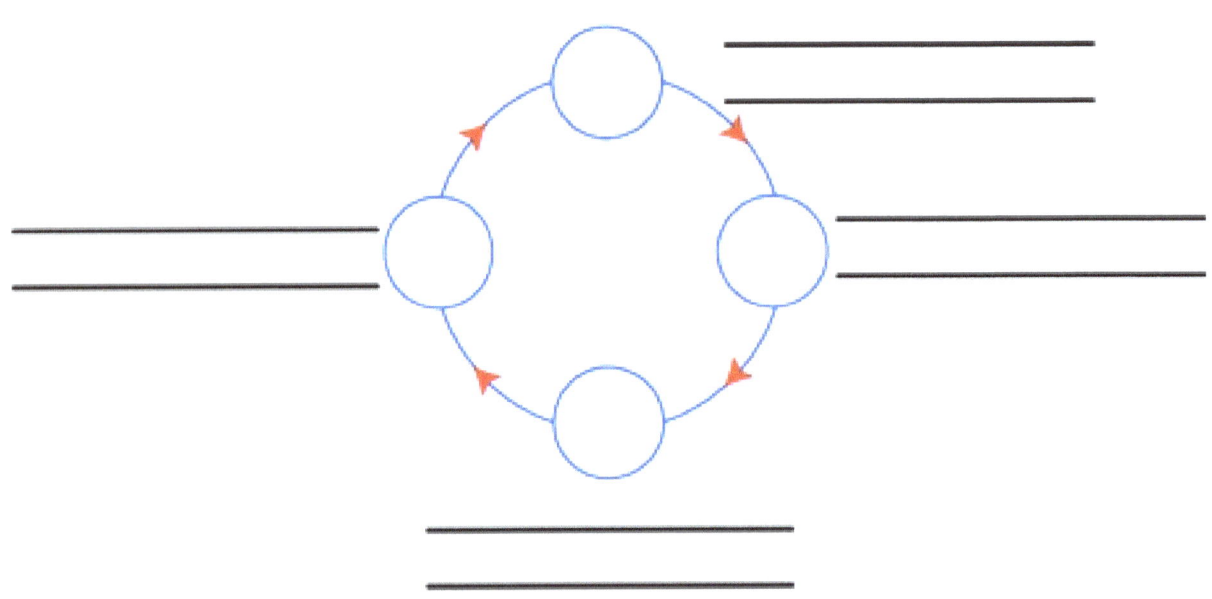

adult

hair turns white

stops drinking milk

eyes are closed

fully grown

old adult

puppy

newborn

©2013 OnBoard Academics, Inc.

Complete the dragonfly's life cycle.

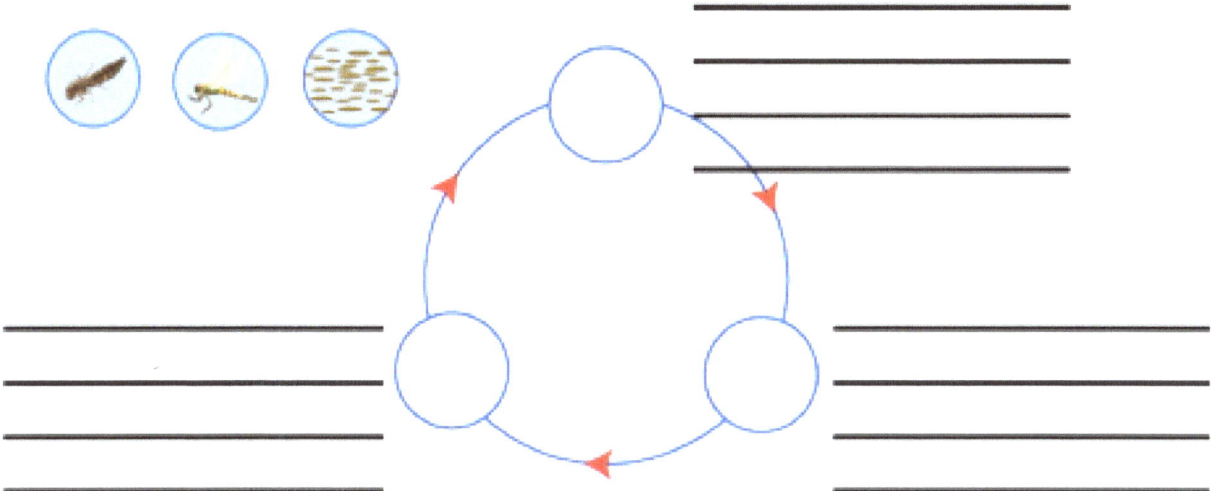

about four years
larva
lives in water

lives in air and land
adult
lives in still water
about two months
egg

also called nymph
up to a few months
may hatch in 5 days
has wings

Do you know your infant animal names?

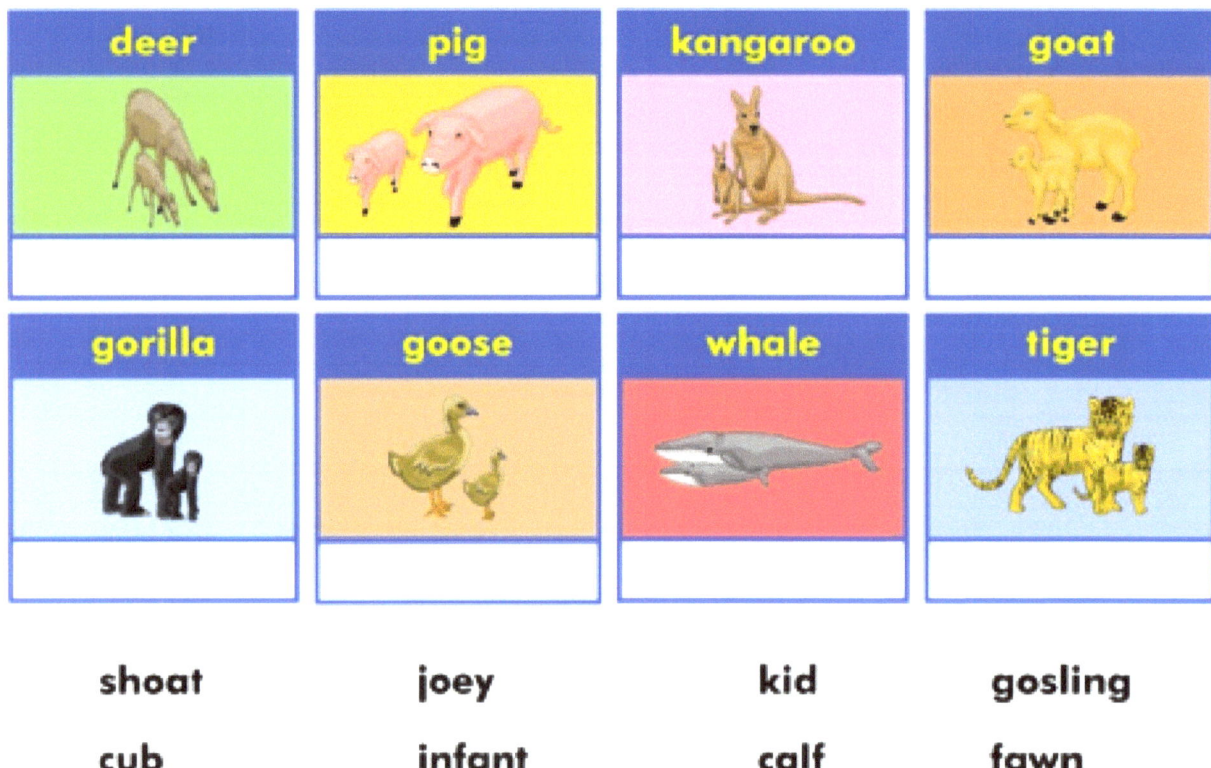

deer	pig	kangaroo	goat

gorilla	goose	whale	tiger

shoat   joey   kid   gosling

cub   infant   calf   fawn

What is my average life span?

Can you guess at the average life span of these animals. Insert the first two letters of the animals name into the circle that you believe represents its life span.

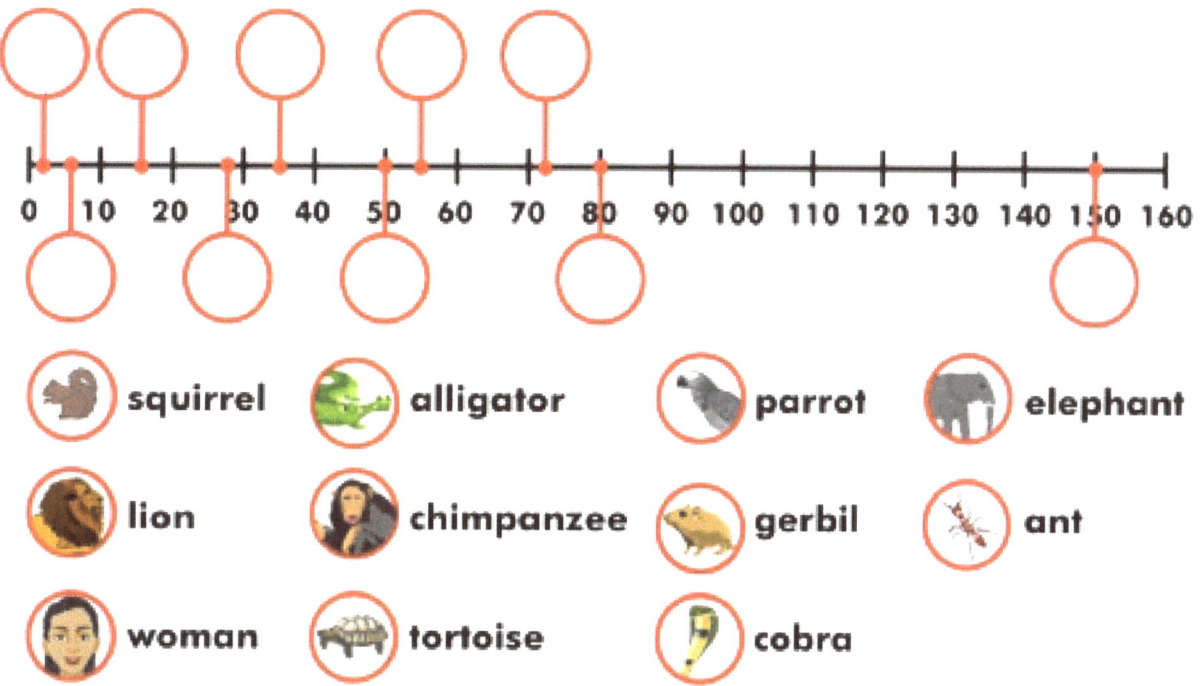

🐿 squirrel	🐊 alligator

All animals eventually die; this is the natural end of a life cycle. What factors do you think influence how long an animal lives on average?

answer

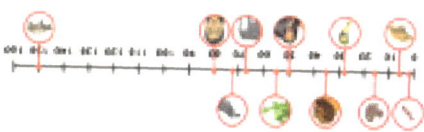

Name: _____

Animal Life Cycle Quiz

1. The set of stages that a living creature goes through is called a _____.
   a. child cycle
   b. life cycle
   c. human cycle

2. How many days does an embryo take to grow into a chick
   a. 12
   b. 22
   c. 21

3. Female chickens are called roosters. True or false?

4. A puppy turns into a full grown dog in about _____.
   a. 1 month
   b. 6 months
   c. 24 months

5. The second stage of a dragonfly's life cycle is called _____.

6. A dragonfly spends most of its life as a larva. True or false?

Metamorphosis

Metamorphosis

During a human's life cycle we maintain the same basic shape, live in the same environment and eat the generally same types of foods.

Animals that undergo a process called metamorphosis experience great changes during their life cycle. To illustrate this, let's take a look at the life cycle of a butterfly.

A butterfly starts its life cycle as a tiny egg that an adult butterfly normally lays on the leaf of a plant. These leaves act as a food source for when the eggs hatch. In about one to two weeks a larva emerges from the egg; a wormlike creature common referred to as a caterpillar.

Caterpillars are voracious eaters and shed their skins multiple times to accommodate their rapid growth. The larva then enters what appears to be a resting stage when it is called a pupa or a chrysalis stage.

Appearances can be deceptive because during this stage the pupa is going through an amazing transformation when the pupa's in which all of its cells are rearranging and rebuilding themselves. This process takes about a week and leads to the final adult stage when it emerges as an adult butterfly.

The butterfly will then look for a mate to renew the life cycle.

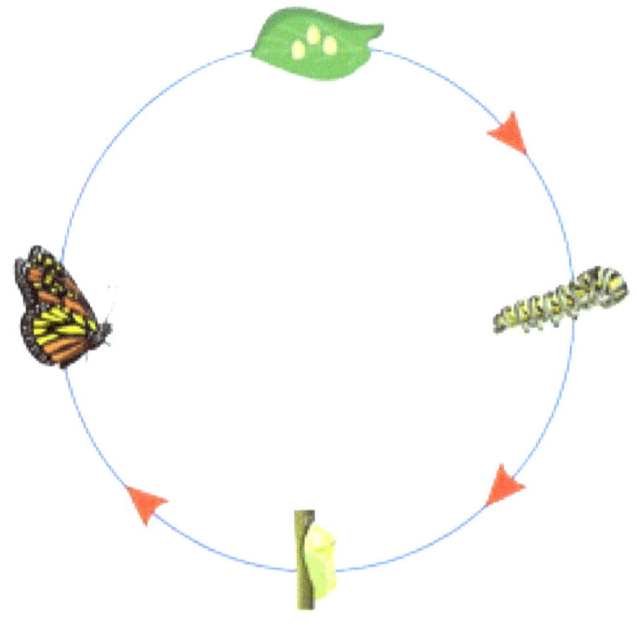

Order the life cycle of a butterfly use both images and descriptions.

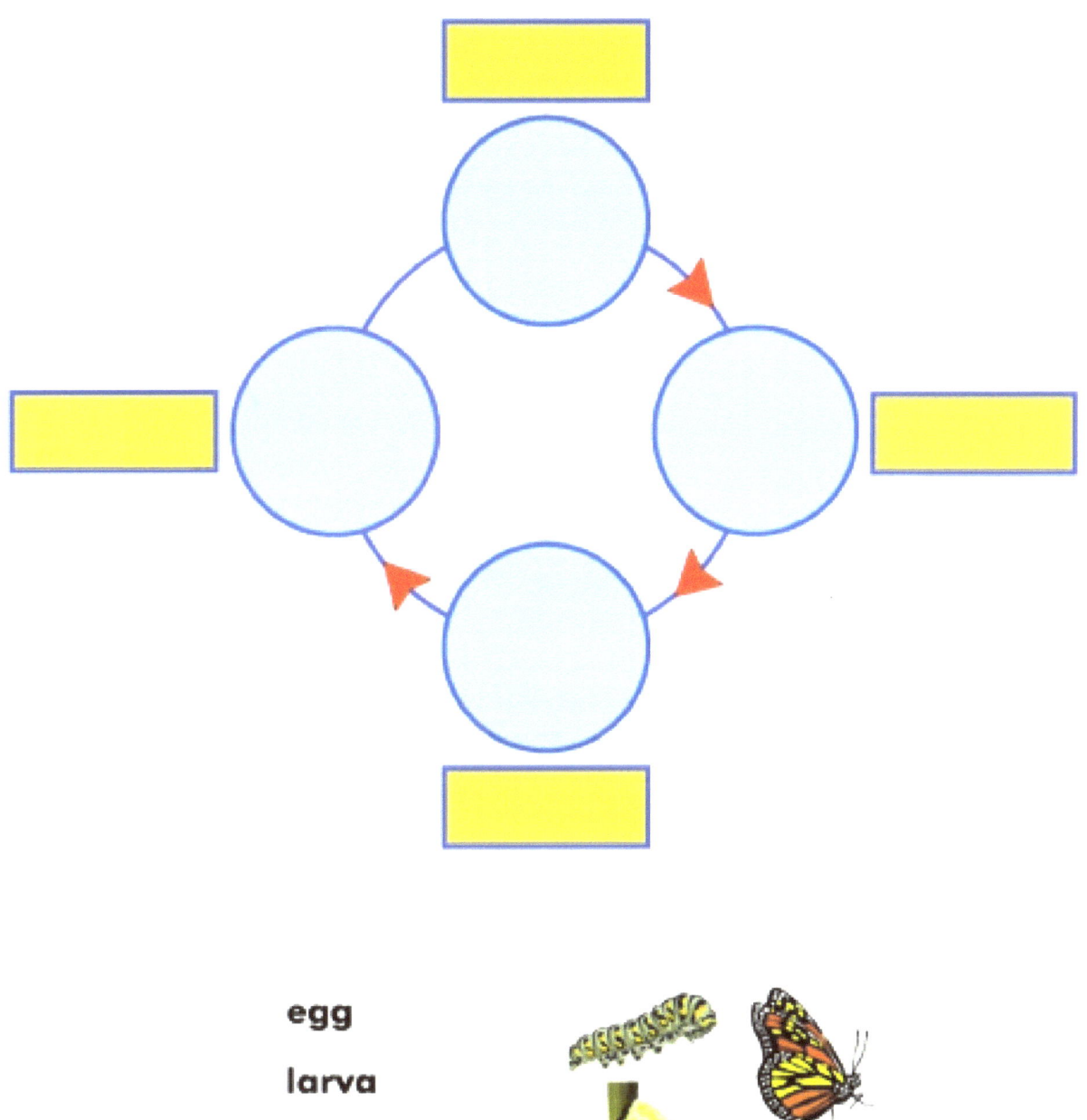

egg

larva

adult

pupa

Moths and butterflies experience a similar process of metamorphosis but other creatures undergo a completely different process of metamorphosis.

Let's look at the life cycle of a frog to understand its metamorphosis.

Like butterflies, frogs start off as eggs. However, frogs eggs are laid in water rather than on leaves and when the eggs hatch a tadpole emerges instead of a caterpillar. Tadpoles are basically a head and a long tale and look somewhat like fish. Like fish, tadpoles breathe with gills.

After a few weeks, the process of metamorphosis begins. The tadpole will develop front legs and the back legs. Its tail will get shorter and it will lose its gills and develop lungs in order to breathe.

When metamorphosis is complete, the adult frog is fully formed.

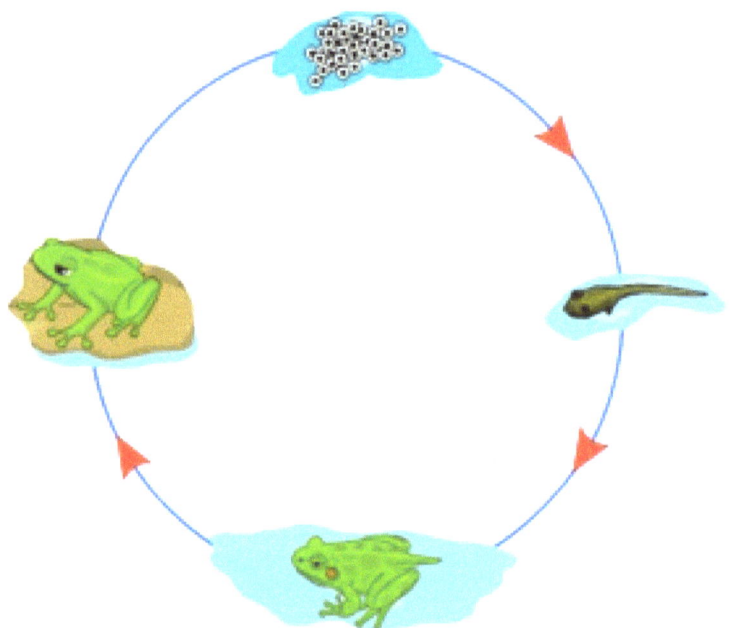

The life cycle of a frog has three distinct stages: egg, tadpole, and adult. At the end of the tadpole stage, the frog undergoes metamorphosis and completely changes its body.

Order the life cycle of a frog.

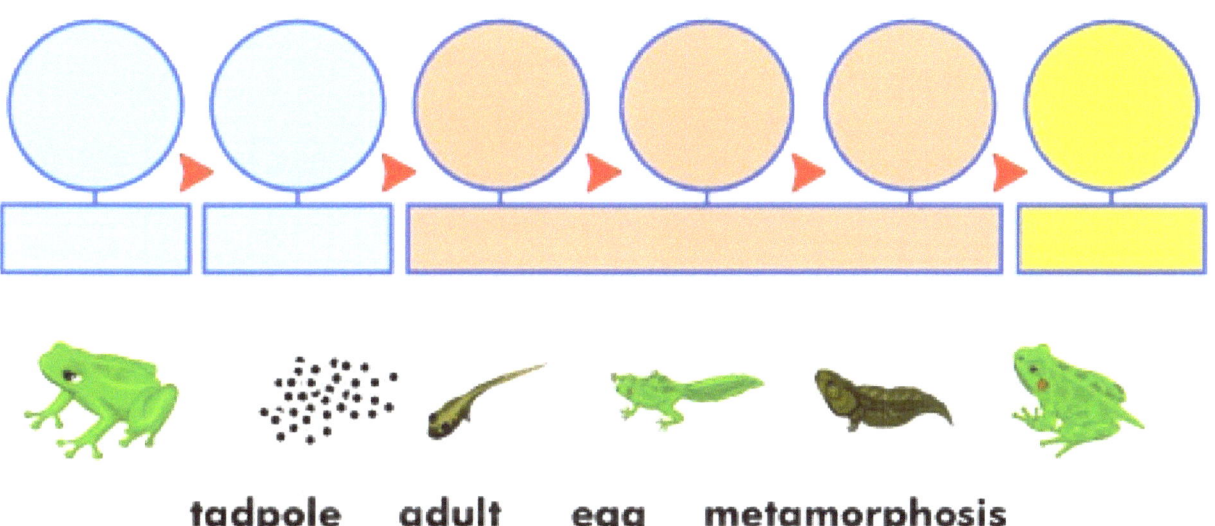

tadpole   adult   egg   metamorphosis

Sort the terms that describe these animals before and after metamorphosis.

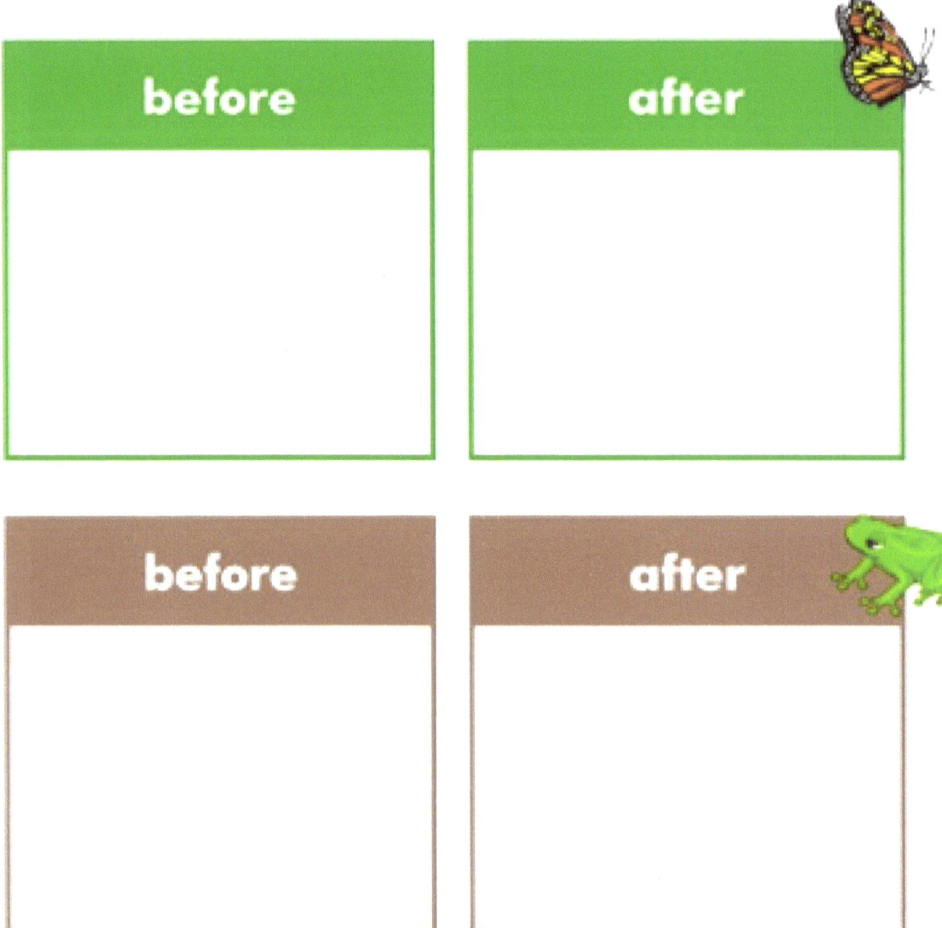

before	after

before	after

has lungs   can only walk

worm-like   grows

eats plants   no legs

can walk and fly   four legs

eats animals   has gills

doesn't grow   has wings

Which of these animals are metamorphic?

√ for metamorphic
X fo not metamorphic

fly ☐ swan ☐ dragonfly ☐ chicken ☐

mosquito ☐ shark ☐ ladybug ☐ salamander ☐

answer

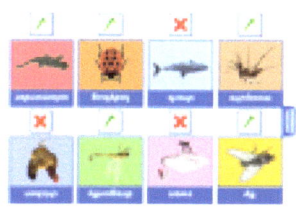

Match these infants with their parents.

fly	ladybug	dragonfly	mosquito	salamander

answer

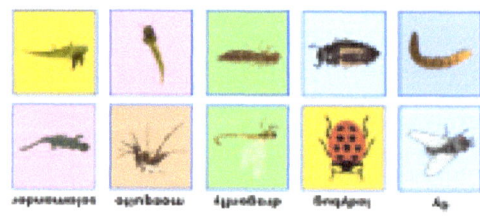

Metamorphosis Quiz

1. The big changes undergone by an animal during its life cycle are called revolution. True or false?

2. Butterflies start their life cycle as _____.
   a. larva
   b. eggs
   c. pupa

3. A pupa is also called a caterpillar. True or false?

4. A larva enters a resting stage called the chrysalis stage. True or false?

5. The eggs of a frog hatch into a _____.
   a. tadpole
   b. fish
   c. frog

6. Like fish, tadpoles breathe with gills. True or false?

7. Tadpoles have a head and four legs. True or false?

Newburyport, MA 01950

1-800-596-3175

OnBoard Academics employs teachers to make lessons for teachers! We create and publish a wide range of aligned lessons in math, science and ELA for use on most EdTech devices including whiteboard, tablets, computers and pdfs for printing.

All of our lessons are aligned to the common core, the Next Generation Science Standards and all state standards.

If you like our products please visit our website for information on individual lessons, teachers licenses, building licenses, district licenses and subscriptions.

Thank you for using OnBoard Academic products.

www.ingramcontent.com/pod-product-compliance
Lightning Source LLC
Chambersburg PA
CBHW050817180526

45159CB00004B/1698